A Path to Universal Basic Income

Michael Kim

First Edition

Contents

Author's Note

Before you read any further, it's worth being honest about what this book is and, just as importantly, what it isn't.

This is not a piece of legislation. It is not a finished economic model, and it is not a claim that the specific numbers, formulas, and mechanisms described in the chapters ahead are ready to be implemented tomorrow, or even that they are the right numbers, formulas, and mechanisms at all. Anyone looking for a fully engineered policy proposal, tested against every edge case and defended against every possible objection, will not find that here. That was never the goal.

This book is a starting point. It is one attempt, by one person, to sketch out what a serious response to a fast-approaching future might look like, in enough detail that it can actually be argued with, rather than staying vague enough to be agreed with by everyone and challenged by no one. Vague ideas are comfortable. They also don't hold up to the kind of scrutiny a policy this consequential deserves. So this book takes a risk that a more cautious version of itself would avoid: it commits to specifics. A specific floor. A specific funding mechanism. A specific approach to children, disability, retirement, and work. Not because those specifics are guaranteed to be correct, but because a concrete proposal is something people can actually respond to, improve, or dismantle piece by piece, in a way that a collection of good intentions never could be.

The conversation about automation, artificial intelligence, and quantum computing, and what they will do to work has, for the most part, stayed inside fairly narrow circles: economists, technologists, policy researchers, and the occasional politician borrowing the idea for a campaign. That's a problem. The changes described in the opening chapters of this book are not going to stay confined to any one profession or industry, and the decisions about how society responds to them shouldn't stay confined to any one circle either. Everyone who works for a living, or hopes to, or is raising a child who will grow up into whatever economy is left standing, has a stake in this conversation. This book is an invitation into it, not a lecture aimed at people who are already convinced.

That invitation is meant especially for the people best positioned to improve on what's written here: economists who can pressure-test the funding model in Chapter 4 far more rigorously than a single author writing alone ever could; labor market researchers who understand the pace of automation better than any single projection can capture; policy experts who know where proposals like this tend to break down in practice; and people from every walk of life whose lived experience of work, scarcity, caregiving, and security is worth more than any statistic in these pages. If you fall into any of those categories and you finish this book thinking a particular chapter gets something wrong, that reaction is not a failure of the book. It's the whole point of writing it this way. Disagreement, sharpened into a better argument, is a far more useful outcome than polite agreement that never gets tested.

None of this is meant to lower the stakes of what's being discussed. If anything, the opposite is true. Chapter 1 lays out the case that the pace of technological change is accelerating faster than most institutions, laws, or individuals can absorb. That case doesn't get weaker just because this book's specific answer to it might need revision. It means the conversation needs to start now, seriously and in public, rather than waiting for a perfect proposal that arrives only once the transition it's meant to address has already happened.

So take what follows in the spirit it's offered: not as a finished blueprint, but as a genuine attempt to think out loud about a future that is coming whether or not society is ready for it, written down in enough detail to be useful to argue with. If it succeeds at nothing else, this book will have succeeded if it gets more people, with more expertise and more perspectives than any one author could bring, talking seriously about what comes next.

Chapter 1

AI, Automation, and Quantum: The Case for Universal Basic Income (UBI)

Every generation believes it lives through unprecedented change. While this is true, the magnitude of change has been accelerating with each generation. The evidence for that claim is no longer philosophical. It is measurable, and it is accumulating faster than most institutions, economies, or individual human beings can absorb.

1.1 A New Kind of Change

For most of human history, technological change was so slow that it was nearly invisible within a single lifetime. A farmer born in the year 1200 could reasonably expect to work the same fields, use the same tools, and live under the same basic economic arrangements as his grandfather. Historians of the pre-industrial world describe economies defined by *limited production, an extreme agricultural economy, and a rigid division of labor* that persisted, largely unchanged, across dozens of generations.[1]

That stability began to break down with the Industrial Revolution, and it has been breaking down faster ever since. Economic historian Robert Gordon has described the last two and a half centuries as a sequence of three great waves: the first, from roughly 1750 to 1830, brought steam power and the railroad; the second, concentrated in the remarkably short span of 1870 to 1900, brought electricity, the internal combustion engine, and indoor plumbing; the third, beginning around

1960, brought computing and the internet. Each wave arrived closer on the heels of the one before it and we are now living through the front edge of a fourth.

The numbers bear this out. During the peak of the Industrial Revolution, engineers spent thirty years improving the efficiency of the steam engine by a factor of five, a compounded rate of about 5.5 percent per year. Today, that pace would be considered almost stagnant. Moore's Law, the doubling of computing power roughly every eighteen months, represents a rate of improvement about ten times faster than the fastest point of the Industrial Revolution.[2]

AI has pushed the curve even further. The amount of computing power used in the largest AI training runs doubled roughly every 3.4 months between 2012 and 2018, a 300,000-fold increase in six years.[3] One analysis of Fortune 1000 technology investment found that overall technology advancement between 2014 and 2024 occurred somewhere between ten and fifty times faster than it did during the 1990s and 2000s, with AI adoption alone moving roughly one hundred times faster than the internet's early diffusion.[4]

This is the first sentence of the argument I will make: the rate of technological change is no longer just fast. It is compounding, and each new wave builds on top of the digital, financial, and cultural infrastructure left behind by the last one.

1.2 From One Generation to the Next

It is worth sitting with how different this is from nearly all of human experience. A useful way to see it is to compare how long it took different technologies to reach mass adoption. The telephone, patented in 1875, took fifty years to reach 50 million users. Commercial air travel took sixty-four years to reach the same milestone. Television, benefiting from a more built-out consumer economy, did it in twenty-two years.

Then the curve bends sharply. The internet reached 50 million users in about seven years. Facebook did it in four and a half. WhatsApp took three and a half years, and Instagram two and a half.[5]

ChatGPT reached 100 million users in two months, a pace UBS analysts, after twenty years of covering the internet sector, said they had never seen before in a consumer application.[6] It reached its first million users in about five days, compared to two and a half months for Instagram and roughly ten months for Facebook.[7]

Each of these technologies didn't just arrive faster than the one before it; it arrived on top of infrastructure the previous technology had already built. The smartphone made the internet portable. Cloud computing made storage and processing power available on demand, to anyone, anywhere. AI is now being layered on top of both, which is a large part of why it is spreading faster than anything that came before it.

1.3 What Comes After AI

If AI feels disruptive, it may only be the opening act. Quantum computing, a fundamentally different way of processing information, one that can evaluate enormous numbers of possibilities simultaneously rather than one at a time, is widely expected to reshape entire industries once it matures. McKinsey's 2026 Quantum Technology Monitor estimates that quantum computing could generate between $1.3 trillion and $2.7 trillion in economic value worldwide by 2035, touching industries from chemicals and pharmaceuticals to financial services, travel, and logistics.[8]

Where AI has largely transformed how we generate and process language, images, and decisions, quantum computing is expected to transform how we simulate and optimize physical and financial systems: drug discovery, materials science, portfolio risk, supply chains, and cryptography among them. Quantum computing companies generated a little over a billion dollars in revenue globally in 2025; McKinsey projects that could grow more than fourfold by 2028.

1.4 The Human Cost of Keeping Up

None of this acceleration is free. It has to be absorbed by real workers, in real jobs, on a timeline that has nothing to do with how quickly they can retrain. The World Economic Forum's Future of Jobs Report 2025, drawing on surveys of more than a thousand employers representing over 14 million workers, projects that 92 million jobs will be displaced globally by 2030.[9]

McKinsey's research suggests that up to 30 percent of hours worked across the U.S. economy could be automated by 2030, and that roughly 70 percent of the skills employers look for will need to change in that same window.[10] Goldman Sachs has put an even larger number on the table, estimating that 300 million jobs globally are exposed to AI-driven automation in some form.[11]

These are not distant, hypothetical projections. They describe a workforce that is now expected to reskill, relocate, and reinvent itself on a timeline measured in a handful of years, again and again, for the rest of a working life. Human beings have never been asked to adapt at this pace before, and there is little evidence that most of our institutions are built to support it.

1.5 Laws Built for a Slower World

If individuals are struggling to keep pace, the institutions meant to protect them are struggling more. Legal scholar Gary Marchant coined the term "the pacing problem" nearly two decades ago to describe the widening gap between technology, which advances exponentially, and law, which moves through legislative and judicial processes built for a far slower era.[12]

The gap has only widened since. More than three years after ChatGPT's public release, comprehensive AI legislation in the United States remains sparse, with regulation scattered across executive orders, agency guidance, and non-binding standards rather than a single coherent framework.[13] One review of AI ethics research pointed out that some of the fair lending laws still being adapted to handle AI-driven decisions are nearly fifty years old, written for a world of paper applications and human loan officers, not algorithms.[14]

This lag matters more than it might first appear. When regulation cannot keep pace with the technologies reshaping the labor market, the costs of that transition (job loss, wage stagnation, and geographic and industry mismatch between old skills and new demand) land almost entirely on workers and families, with little institutional cushion to absorb the shock.

1.6 Why This Changes the Argument for Universal Basic Income (UBI)

Put these threads together and a pattern emerges. Technological change is not simply fast; it is accelerating, wave after wave, each one arriving faster than the last and building directly on the infrastructure of the one before it. The jobs and skills that generation after generation could once count on to remain stable for a working lifetime are now being reshaped every few years. And the laws and institutions meant to soften that transition are, by their own design, built to move far more slowly than the technology they are meant to govern.

This is the gap Universal Basic Income (UBI) is meant to address: not a philosophical preference for redistributing wealth, but a practical response to a labor market that is now changing faster than any workforce in human history has ever been asked to adapt. The chapters that follow will make the case for what a basic income could look like, how it might be funded, and why in an economy being reshaped simultaneously by AI and quantum computing, it may no longer be a question of ideology, but of necessity.

Chapter 2

Basic Existence Without Fear

Chapter 1 made the case that the pace of technological change has outrun the institutions meant to manage it. The labor market is now being reshaped faster than any workforce in history has been asked to adapt, while law and policy remain structurally incapable of keeping up. If that argument holds, the next question is unavoidable: reshaped into what, exactly, and what does a society owe the people caught in the gap?

This chapter answers that question with a definition. UBI is enough money to exist without fear. It is not enough to thrive. It is not enough to live in comfort. It is enough to exist without constant anxiety about survival. It means having enough to eat, somewhere safe to sleep, a way to stay connected to the world, and the ability to participate in society without the constant threat of losing any of it.

That distinction between existence and prosperity is the foundation for everything else in this book. Get it wrong in either direction and the whole policy collapses: too little, and UBI becomes a hollow gesture that still leaves people choosing between rent and medicine; too much, and it becomes politically and fiscally impossible, inviting exactly the kind of backlash that has killed basic income proposals before. So before I discuss funding, work incentives, or implementation, I must define the floor with precision.

2.1 A Simple Test for What Belongs in the Floor

Every proposed UBI expense in this book is run through the same test, applied in both directions:

- Does this expense prevent a person from meeting a basic human

need > or participating in society? If yes, it potentially belongs in the > UBI calculation.
- Does this expense merely improve someone's comfort, status, > convenience, or luxury? If yes, it does not belong in the UBI > calculation.

This test is deliberately blunt. It is not asking what would make someone's life better, easier, or more enjoyable; almost anything would. It is asking what a person cannot go without and still function as a participant in modern society. That second qualifier, participation in society, matters as much as survival itself. A UBI calculated only on caloric and shelter minimums, the old-fashioned subsistence standard, would guarantee biological survival while still locking people out of employment, education, and civic life. Modern minimum-income research has converged on this broader standard for exactly that reason.

The United Kingdom's Minimum Income Standard project, one of the most rigorous ongoing efforts to define this floor, defines it as "a minimum standard of living that is more than just food, clothes and shelter. It is about having what is needed to have the opportunities and choices necessary to participate in society."[15] That project builds its budgets from detailed public deliberation including groups of ordinary citizens negotiating, line by line, what a household actually needs, rather than from a government formula alone, and it is recalculated every year against current prices.[16]

2.2 What the Floor Includes

Applying the test, UBI should cover:

- Food
- Shelter
- Clothing
- Basic communication: phone and internet service
- Basic utilities: water, electricity, gas, and internet
- Basic transportation
- Education through university

- Health care, if universal health care is not already in place by the > time UBI is implemented

After these categories are covered, the recipient should have some amount of discretionary income left over; not much, but enough to make small choices that are entirely their own. A floor with zero discretionary income isn't dignity, it's a ration card. The test above is what keeps that discretionary margin from ballooning into something the floor was never meant to fund.

2.3 A Flat Amount, Nationally

The UBI amount should be flat, the same for every adult recipient, regardless of demographic group or region. This is a deliberate departure from how most cost-of-living adjustments work today, and it deserves a direct defense.

The obvious objection is that cities cost more to live in than rural areas, so a flat national number will be tight in Manhattan and generous in rural Mississippi. That's true, and it's also not a flaw. It's the point. UBI is not designed to guarantee the median cost of living in every zip code a person might choose to live in. It guarantees basic existence, calculated as a national average. If someone chooses to live in an expensive city, UBI guarantees their basic existence there, not their ability to afford that city's median apartment.

And as I'll discuss later, I propose that the shift to UBI be gradual, matching the pace and acceleration of AI, automation, and quantum productivity gains. Over time, if people don't need to live in a big city for work, they will be much more likely to move to areas where the cost of living is lower as we witnessed in 2020.

2.4 How the Amount Is Calculated and Adjusted

The UBI floor should be calculated using a defined basic-needs methodology such as a basic needs index, though the exact instrument matters less than the discipline of using one consistently. The calculation should be averaged across the entire country, and recalculated every year. Minimum Income Standard researchers in the UK follow this model already: budgets are rebuilt from current prices annually, capturing inflation in essentials like food, energy, and rent rather than relying on a general inflation figure that can understate what's happening to the specific goods low-income households depend on most.

This distinction matters. Research into the U.S. Consumer Price Index found that the cost of a basic-needs budget for low-income families rose faster than the general CPI over the periods studied, meaning a UBI indexed to general inflation alone could quietly lose purchasing power against the actual necessities it's meant to cover.[19] A dedicated basic-needs index, rebuilt annually, avoids that slow erosion.

The flat amount remains the same at retirement age. UBI is not a pension and does not attempt to replace one. It is the same floor at every adult age, on the theory that the need for basic existence doesn't change because a birthday has passed. UBI would gradually expand Social Security with my proposal. Retirement savings or other pension systems can and should sit on top of this floor for those who have them; UBI simply guarantees the floor doesn't disappear for those who don't.

2.5 Children

UBI for children should not be a flat, adult-sized amount. It should be calculated differently, by age, using the same basic-needs logic applied to the actual cost of raising a child at each stage of development.

Existing research already does much of this work. The U.S. Department of Agriculture's "Expenditures on Children by Families" study, the most detailed government breakdown of child-rearing costs ever produced, even though it hasn't been updated since 2015, allocated household spending to children by category: housing calculated as the value of an additional bedroom, food from federal food-plan budgets, health care from national medical expenditure surveys, clothing, and education from direct child-specific spending data.[20] That methodology, updated to current prices and rebuilt on the same basic-needs logic as the adult floor, is a reasonable starting model for a child UBI calculation, adjusted by age band, since costs shift substantially from infancy through the teenage years.[21]

A family's UBI for children should also come with diminishing returns as family size grows. The reasoning is not a judgment about family size; it's a recognition that many of the largest costs of raising children such as housing space, utilities, transportation, even food purchased in bulk do not scale linearly per child the way they do for a first child entering a household with none. A UBI designed to fund basic existence, not to subsidize family size decisions indefinitely, should reflect that.

2.6 Disability and Special Needs

People with disabilities and other special needs face real, well-documented costs that a standard adult or child floor will not cover, and UBI should account for that directly through a separate, additional calculation, built the same way as the child model, using existing cost-of-disability research as a starting point rather than reinventing the methodology from scratch.

The UK charity Scope has published the most detailed ongoing research on this question, calculating what it calls the "Disability Price Tag," the additional amount a disabled household needs to reach the same standard of living as a non-disabled household. Their most recent figures put that gap at roughly £975 to over £1,000 a month for the average disabled household, even after accounting for existing disability benefits.[22] The methodology compares a disabled household's ability to afford the same standard-of-living basket as a non-disabled household, rather than simply itemizing medical costs.[23]

A U.S. equivalent, built the same way and recalculated annually alongside the base UBI floor, would let disability-related UBI supplements track real costs rather than a fixed, one-time-legislated amount that goes stale within a few years.

2.7 What the Floor Does Not Cover

Running the same test in reverse, some examples of what UBI should not fund:

- Child care
- A luxurious lifestyle
- Expensive personal hobbies
- Anything for non-citizens
- Expensive vacations, unless funded by a person's own savings over > time
- Expensive technology, unless funded by a person's own savings
- A mortgage on a home, unless funded by a family's own savings and > pooled resources
- Frequent restaurant meals
- More expensive cars
- Pets
- Other people

That last item deserves emphasis, because it's the one most likely to generate confusion. A UBI is intended to guarantee one person's basic existence. It is not automatically an obligation to financially support other adults no matter how close that relationship is. Two adults living together each receive their own individual UBI; neither is entitled to a larger amount because the other has chosen not to work, and neither is required to share their floor with someone who isn't covered by one. UBI supports the individual. It does not underwrite every financial responsibility that an individual voluntarily takes on.

Put together, this is the standard: UBI should guarantee the material conditions necessary for a person to participate in society without being threatened by hunger, homelessness, untreated illness, or an inability to meet basic human needs. UBI does not prevent anyone from becoming wealthy, and it does not stop anyone from earning more. That possibility, and how it should interact with the floor, is the subject of the next chapter.

2.8 The Ability to Say No

A floor calculated this way does something that no amount of means-tested, conditional, or employer-dependent assistance can do: it gives a person the standing to walk away. Not from work in general, but from any specific arrangement that has become intolerable. With a guaranteed floor under them, a person can say no to:

- An abusive employer
- An unsafe job
- An exploitative relationship
- An intolerable living situation
- Working seven days a week simply to survive

This is not a hypothetical benefit. Canada's Manitoba "Mincome" guaranteed-income experiment in the 1970s, one of the most closely studied basic income trials on record, found that payments increased some measures of wives' bargaining power within their households and lowered conflict specifically related to financial strain, consistent with the theory that a guaranteed income gives a financially dependent partner a genuine exit option from a bad relationship.[24] A 2008–2009 basic income pilot in Namibia found a similar pattern: unconditional transfers appeared to reduce young women's economic dependence on men for survival.[25] Reviewing this and related research, one analysis of UBI's effect on the modern employment relationship concluded that a basic income lets workers reject exploitative or oppressive jobs and strengthens their bargaining position simply by giving them a credible option to walk away.[26]

This is the deeper purpose behind everything in this chapter. A UBI floor is not a reward for existing, and it is not an experiment in reshaping how people live. It's insurance against the specific kind of powerlessness that comes from having no alternative, the powerlessness of needing this job, this landlord, or this relationship regardless of what it costs, because the alternative is nothing at all. The next chapter turns to what happens above that floor: how UBI should interact with work, and why guaranteeing survival is not the same as removing the incentive to build something more than survival.

Chapter 3

UBI, Future Work, and Inflation

Two objections come up faster than any others whenever UBI is discussed. The first: won't people just stop working? The second: won't printing that much money just cause runaway inflation? Both are fair questions, and both rest on a misunderstanding of what the system actually requires. This chapter takes them in order, then lays out a concrete path for getting from where we are today to where I think we are going.

3.1 The System Doesn't Need Everyone to Work

Start with a premise that sounds radical but is really just an observation: the economy does not require every adult to work. It requires enough people to produce the goods and services society demands. Those are very different thresholds, and the gap between them has been widening for two centuries, as automation has let fewer people produce more, the same trend Chapter 1 described accelerating with AI and, soon, quantum computing. UBI doesn't have to change that trajectory. It only has to change who is protected while it happens.

3.2 Work Still Pays, The Equation Just Shifts

Crucially, everyone who works still receives their UBI. It is not a replacement for earned income, and it is not clawed back as wages rise. That single design choice reframes the entire relationship between working and surviving:

	Today	With UBI
Work →	survival + additional prosperity	additional prosperity
UBI →	(not applicable)	survival

Today, most people work primarily because they have to. Survival is bundled with the job. Under UBI, survival is unbundled from employment. Work no longer has to cover both; it only has to cover the second half of the equation. That still leaves a large, straightforward economic incentive to work, because UBI by design is calculated at the basic floor, not at a comfortable standard of living.

Someone receiving UBI might still want any of the following, none of which the floor is designed to provide:

- A larger house
- A better car
- Vacations
- Expensive hobbies
- Savings
- Retirement wealth
- Status
- Achievement
- Social interaction
- Meaningful work
- Entrepreneurship
- Additional security

Every one of those still requires income UBI doesn't provide. That is the incentive structure that keeps supply flowing even after the floor is guaranteed; not fear of destitution, but the ordinary human desire for more than bare survival.

3.3 A Gradual Shift, Not a Cliff

It's also worth being precise about the pace of change this chapter assumes. Companies will need fewer workers over time to produce the same output. That trend is not in dispute, but the shift will most likely be gradual, not sudden, and employees will still be needed throughout most of the transition. Labor economists Daron Acemoglu and Pascual Restrepo describe this as a race between two forces: a displacement effect, where machines take over tasks people used to do, and a reinstatement effect, where the same technological change creates new tasks only people can do. Historically, the reinstatement effect has tended to offset the displacement effect and stabilize the labor market over time, rather than eliminating work outright.[28]

That balance is not guaranteed to hold forever, and there's real evidence it is already straining. A 2026 analysis found that task displacement has intensified over the past four decades while the creation of new work has not always kept pace, with early signs that the most recent wave of AI-driven automation disproportionately affects entry-level workers.[29] This is precisely why UBI matters as a floor under the transition rather than a replacement for it: nobody can predict exactly how quickly automation will eliminate any given job. The honest approach is to build a system that can be adjusted one year at a time as the real data comes in, rather than one that assumes either a soft landing or a sudden collapse.

3.4 Does UBI Cause Runaway Inflation?

It could, in theory. Inflation is not a fixed penalty attached to any policy that puts money in people's hands; it is a relationship between additional demand and the economy's capacity to produce goods and services. If demand rises faster than supply can follow, prices rise to ration the difference. If supply can absorb the new demand through existing slack, productivity gains, or a gradual enough rollout, prices don't have to move much at all. UBI's inflation risk is not fixed; it's a design and implementation question.

The best real-world evidence available comes from Alaska, which has paid every resident an unconditional annual dividend from the Alaska Permanent Fund since 1982, the closest thing to a long-running, statewide UBI experiment in the United States. A synthetic-control study comparing Anchorage's Consumer Price Index to a matched set of other U.S. metro areas after the dividend began found no meaningful difference in CPI attributable to the policy.[30] A separate long-run comparison of Alaska's CPI to the national CPI found that Alaska's inflation actually ran lower than the rest of the country after the dividend was introduced, having tracked roughly equal to the national rate before it. This is suggestive evidence that a sustained unconditional cash program does not have to produce runaway inflation.[31]

Labor-supply research on the same dividend backs this up from another angle. Using Alaska as a natural experiment, economists found no significant overall effect on employment from the payments, with some studies finding a modest increase in part-time work rather than an exodus from the labor force.[32] If a sustained, unconditional cash payment neither drove Alaskans out of the workforce nor pushed prices up faster than the rest of the country, that is meaningful evidence against the assumption that any UBI must be inflationary by default.

None of this means a national UBI is risk-free. Alaska's dividend has historically been a few thousand dollars a year, smaller than a full basic-needs floor, and its funding source (oil royalties) doesn't map directly onto how a national program would be financed. But it does mean the inflation objection is an engineering problem, not a law of economics. And the biggest lever available for managing that engineering problem is the pace of the rollout itself.

3.5 Implementation: Build on Social Security, Don't Replace It

The proposal I make is to implement UBI by expanding the existing Social Security system rather than building a parallel one from scratch. People who have already paid into Social Security would continue receiving those earned benefits in full, on top of their UBI once the program begins. This is not a redundancy; it's a fairness mechanism.

Someone who spent forty years paying into a system on the promise of a specific benefit shouldn't see that promise diluted or replaced by a flat universal payment calculated on a different basis. They receive both.

This design choice does double duty. It's more equitable for people who already paid into the system, and it naturally increases the rate at which older workers opt out of the labor force as the program expands, which is also, not coincidentally, the mechanism for controlling inflation during the rollout.

3.6 The Retirement-Age Lever

Rather than launching a full national UBI for every adult on day one, the rollout should begin by gradually lowering the Social Security eligibility age, expanding the pool of people receiving both their earned benefit and UBI a little further each year.

There is direct precedent for phasing a Social Security age change in gradually rather than all at once. The 1983 Social Security Amendments raised the full retirement age from 65 to 67, but rather than imposing the change immediately, Congress phased it in over more than three decades, tied to birth year, so the system and the workforce could adjust incrementally.[33] That same mechanism, a phased incremental age adjustment rather than a single cutover, can run in reverse: instead of raising the age to reduce costs, gradually lower it to expand coverage, adjusting the pace based on real conditions rather than a fixed timetable set decades in advance.

One specific dial for setting that pace could be the national unemployment rate. Not total unemployment, but the standard measure of people actively looking for work who can't find it. The Federal Reserve already manages monetary policy around a similar concept: its dual mandate targets maximum employment alongside stable prices, and Fed officials have historically estimated the "natural rate" of unemployment consistent with that goal at somewhere around 4 to 4.5 percent, adjusting their target as labor-market conditions evolve rather than fixing it permanently.[34] As a starting point, I propose borrowing that same logic for the Social Security age lever: track unemployment every year, and if it is running high, lower the Social Security eligibility age to pull people out of the labor force and reduce competition for scarce jobs. If unemployment is low, hold the age where it is. The age should never move in the other direction. Once lowered, it stays lowered, and the endpoint of the whole mechanism is the point at which the eligibility age has come down far enough that every adult in the country is effectively covered by UBI.

This turns the retirement age into the primary lever for controlling both unemployment and the pace of the UBI rollout at the same time. It ties the policy directly to observable, yearly data instead of a guess about how fast automation will eliminate jobs. And because the shift happens gradually, a year at a time, it gives policymakers room to watch the relationship between new demand and productive capacity described above, and adjust the pace before an inflation problem, rather than after one.

3.7 Letting the Market Point the Way

A gradual transition also creates time and information that a sudden one wouldn't: a chance to see, industry by industry, where labor is still needed and where it isn't. The market already signals this constantly through wages, job openings, and shortages. The task for policy is to notice those signals and make the industries they point to more attractive, rather than leaving that adjustment to chance.

For industries anticipated to remain in high demand even as automation advances, government and industry already have a working playbook worth expanding:

- Free or subsidized education for high-demand fields
- Apprenticeships that combine paid work with training
- Higher wages in shortage occupations
- Relocation assistance to underserved regions
- Faster certification pathways

- Housing assistance tied to service in shortage areas

These aren't hypothetical tools. Federal programs like the Nurse Corps Scholarship and Loan Repayment program already pay nursing students' tuition or repay their debt in exchange for service in facilities facing critical shortages, and several states run parallel loan-forgiveness and tuition-assistance programs of their own for the same purpose.[35] Hospital-run nursing apprenticeship programs which pay apprentices a wage while they train, cover tuition costs, and guarantee a job on completion, have shown early success getting people into nursing careers who might not otherwise have entered the field, precisely by removing the financial barriers to entry rather than simply raising the pay ceiling at the top.[36]

This is the model that should scale to every industry the labor market signals it needs, as automation continues to reduce demand elsewhere: let wages and shortages identify where people are needed, then use these same tools to make it easier for people to move into those roles.

3.8 Putting It Together

None of this eliminates uncertainty. Nobody can say today exactly how fast automation will reduce labor demand, exactly how large the resulting unemployment pressure will be, or exactly how quickly the Social Security age can safely come down. But that uncertainty is precisely the argument for a gradual, adjustable rollout rather than a single, fixed program set in stone at launch. Expand Social Security first, so people who've paid in are treated fairly. Use the unemployment rate as the yearly dial for lowering the eligibility age, following the same phased-implementation model Social Security has already used once before. Let the labor market's own signals, not a central planner's guess, determine which industries get targeted support to stay staffed. And because the transition happens gradually, policymakers get to watch the relationship between new demand and the economy's productive capacity in real time, and adjust the pace before problems compound rather than after.

UBI does not require the end of work, and it does not require betting the whole economy on a single irreversible policy change. It requires a floor that grows in step with how quickly the economy can actually support it, which is exactly the kind of gradual, data-driven system the uncertainty described in Chapter 1 demands.

Chapter 4

My Proposed Funding Model

4.1 The Coming Revenue Problem

If the labor force is expected to shrink relative to the size of the economy, then revenue from income and payroll taxes will shrink with it. Payroll taxes are, definitionally, a tax on wages. Fewer people working, or a smaller share of national income going to wages at all, means a shrinking base for the tax system that currently funds both general government and Social Security specifically.[37]

That shrinkage is not a future hypothetical; it's already showing up in the data. Labor's share of U.S. economic output — the portion of GDP that flows to workers as wages and salaries — fell to 53.8 percent in the third quarter of 2025, the lowest level recorded since the Bureau of Labor Statistics began tracking the figure in 1947.[38] Over the same stretch, corporate profits as a share of GDP climbed to roughly 15.85 percent, nearly double the 8 percent share they represented in 1982, while employee compensation's share fell from 66.6 percent to 61.9 percent of GDP over that same 44-year window.[39]

Social Security itself is already feeling this squeeze, well before AI-driven automation has meaningfully entered the picture. The program's trustees project that its combined trust funds will be unable to pay full scheduled benefits starting in 2034, at which point the law requires an automatic, across-the-board cut of roughly 19 to 23 percent unless Congress acts before then.[40] A shrinking pool of wage-earning workers relative to beneficiaries is one of the central drivers of that shortfall and it is a preview, not an aberration, of what a genuinely labor-light economy would do to any tax system built exclusively around wages.

4.2 Why Should Financing Stay Tied to Human Labor?

Put those numbers together and a question becomes hard to avoid: why should the financing of society remain primarily dependent on taxing human labor, at the exact moment the economic system is generating a growing share of its value through capital and technology instead? A tax base built for an economy where most value was created by people doing work makes less and less sense as that description becomes less and less true. If value creation is shifting from labor to capital, the tax base funding society's shared obligations should be allowed to shift with it, rather than continuing to lean ever harder on a shrinking foundation.

This is a core argument of this chapter: companies should contribute to UBI because the economic system is increasingly generating value through capital and technology rather than human labor. Not as a penalty for automating, and not as a punishment for success; simply as a recognition that the tax base should evolve with the economy rather than remaining permanently anchored to wages.

4.3 Not a Tax on "AI"

It's worth being precise about what this chapter is not proposing, because a version of this idea has been publicly debated before and largely rejected in the form it was proposed. In 2017, Bill Gates suggested that companies deploying robots to replace human workers should pay a tax on those robots roughly equivalent to what they'd have paid in income and payroll taxes on a human worker doing the same job.[41] The idea drew sharp criticism, including from economist Lawrence Summers, who argued it amounted to "protectionism against progress" and would shrink the overall economic pie rather than simply redistributing it.[42] A more practical objection proved just as damaging: nobody could cleanly define what counts as a "robot" for tax purposes, especially as automation increasingly lives inside ordinary software rather than a distinct physical machine on a factory floor — which makes a tax targeted at automation itself nearly impossible to administer fairly.

This chapter's proposal sidesteps both problems by not trying to tax automation, AI, or robots as a category at all. The goal isn't to tax "AI." The goal is to gradually shift the tax burden from human labor toward the economic surplus generated by capital and technology, as that surplus becomes a larger share of national production — taxing outcomes (corporate profit and revenue), not the specific mechanism (a robot, an algorithm, a piece of software) that produced them. That distinction avoids the definitional trap that doomed the robot tax, and it avoids the "shrinking the pie" critique as well: this chapter's proposal is explicitly designed to tax economic surplus without discouraging productive capital investment, not to penalize the investment itself.

4.4 The Proposal: Tax Corporate Income Through the Social Security Framework

Concretely, this book proposes that corporations be taxed on their income, and that the most practical mechanism for doing so is to leverage the Social Security framework already in place — the same expansion Chapter 3 proposed as the delivery mechanism for UBI itself. Rather than building an entirely new tax system from scratch, extend the existing one: charge companies an increased Social Security contribution rate, calculated on their profit, their revenue, or some blend of both.

Several design principles should govern how that rate is structured:

- A marginal, progressive system. Use the same structure used for > individual income tax, where the rate rises in steps as the > taxable amount grows, rather than a single flat rate applied to > every dollar.
- Rates adjusted with international competitiveness in mind, so that > companies aren't penalized simply for choosing to operate in the > United States rather than relocating elsewhere.

- A gradual shift of the employee's share of federal payroll taxes > toward corporations over time, mirroring the gradual rollout > described in Chapter 3 rather than a single abrupt change.
- Full elimination of the employee-side Social Security contribution > once UBI is fully implemented. Working people shouldn't be > financially penalized at a personal level for choosing to work, > once the floor Chapter 2 describes is already guaranteed.

4.5 Nonprofits Stay Exempt

Nonprofit organizations should remain exempt from this expanded contribution, exactly as they are exempt from corporate income tax today. The rationale doesn't change simply because the tax mechanism is evolving: nonprofits don't generate the kind of distributable profit surplus this chapter is targeting, and taxing them would undercut the charitable, educational, and community functions they exist to perform; functions that, in many cases, directly reduce the burden UBI itself is designed to carry.

4.6 Profit Is a Component, Not the Whole Formula

One design detail deserves particular emphasis: profit is probably one component of the funding formula, rather than the entire formula. A tax based purely on reported profit creates an obvious loophole for exactly the kind of company this chapter is most concerned with: a highly automated, high-revenue business that minimizes reported profit through aggressive reinvestment, depreciation, or accounting structure, while still generating enormous economic value and needing very few human employees to do it. A blended base, weighing both revenue and profit rather than profit alone, closes that gap and keeps the tax tied to the actual scale of economic activity a company is responsible for, not just the number its accountants report at the bottom of the income statement.

4.7 Why This Is the Right Foundation

None of this requires inventing a new bureaucracy or an entirely new tax regime. It requires expanding a system that already exists, already has the administrative infrastructure to collect contributions and disburse benefits, and is already the natural home for a gradually implemented UBI rollout. Funding UBI through an expanded, progressively structured corporate Social Security contribution does three things at once. First, it ties the tax base to where economic value is actually being generated as that shifts from labor to capital. Second, it avoids the definitional and competitiveness failures that sank the robot tax. Third, it lets the same gradual, data-driven rollout mechanism described in the last chapter govern both sides of the ledger: who receives UBI, and who pays for it at the same pace, adjusted by the same evidence, year by year.

Chapter 5

UBI and Purpose in Life

Chapter 4 answered a practical question: who pays for UBI. This chapter answers a harder one: what happens to a person once survival is no longer in question.

5.1 The Question Money Doesn't Answer

Consider a fact that should complicate any simple story about money and purpose: people who are already well-off, who have more than enough to thrive financially, still regularly ask themselves what they are supposed to do with their lives. This is not a hypothetical concern. It shows up constantly among people who have already achieved financial security through a career, then lose the structure that career provided.

Research following retirees through the Health and Retirement Study, and separate research published in the Journal of Occupational Health Psychology, both point to the same pattern: people whose sense of self was built primarily around their job often experience genuine psychological distress once that job ends, even when their financial needs are fully met.[44] The problem in these cases was never money. It was the sudden absence of a structure that used to answer the question of what to do with one's time and how to see oneself as useful. That is worth sitting with before assuming that guaranteeing survival will leave people adrift. Plenty of people already have their survival guaranteed today, through savings, pensions, or wealth, and the openness of purpose it creates is not despair. It is simply an open question that each person has to answer for themselves.

This chapter argues that UBI does not remove work from people's lives. It removes something narrower and more specific: the assumption that paid employment is the only legitimate form of productive activity a person can engage in.

5.2 Employment Is Not the Same Thing as Work

These two words get used interchangeably, and that habit hides an important distinction. Employment is work performed in exchange for wages. Work is a much broader category, and most of what people spend their lives doing that genuinely matters to them and to the people around them was never wage labor to begin with. People work when they are:

- Raising children
- Caring for parents
- Gardening
- Volunteering
- Creating art
- Building things
- Studying
- Teaching
- Writing
- Coaching
- Participating in community organizations
- Maintaining their home
- Helping neighbors
- Starting businesses
- Inventing things
- Creating music
- Exploring
- Practicing their faith or spiritual life

None of that is idleness. All of it is work, in the plain sense of effort directed toward something that matters. The only thing missing, in most cases, is a paycheck.

5.3 Changing the Math on Risk

Guaranteed survival changes more than how people spend their afternoons. It changes what people are willing to risk. Starting a business, today, often carries a very specific kind of fear behind it: if this fails, I cannot pay my mortgage. Under UBI, that same decision carries a very different calculation: if this fails, I still have food and shelter. That is not a small emotional difference. It is the difference between a decision made from scarcity and a decision made from security, and it changes what kinds of risks people are willing to take.

There is already real evidence for this effect, not just theory. Stockton, California's guaranteed income pilot gave 125 residents 500 dollars a month with no conditions attached. Researchers found that the payments reduced income volatility and, in their words, created new capacity for self-determination, goal setting, and risk taking. Recipients did not withdraw from work; they moved into full-time employment at more than twice the rate of the control group, in part because they finally had the financial cushion to take the risk of a day off to interview for a better job, something that is genuinely dangerous to attempt when every missed shift threatens next month's rent.[46]

Extend that same logic to entrepreneurship specifically, and the case becomes clear. With a survival floor guaranteed, more people are likely to attempt starting a business, pursuing art, launching a nonprofit, inventing something, writing a book, or working as an independent contractor. Some of these attempts may "fail" by current measures because in today's world, they have to secure an individual's survivability. With UBI, it is simply how one chooses to spend their time if that's what they enjoy. It stops meaning the loss of a home or the inability to feed a family. And if the new venture ends up being profitable, that's simply a bonus.

5.4 Reclaiming Time, Not Just Income

A person working a standard full-time schedule frequently does not have time left over to cook healthy meals, exercise regularly, sleep enough, maintain their home, care for aging parents, spend meaningful time with their loved ones, or participate in community life. This is not a personal failure of time management; researchers who study what they call time poverty have found that Americans working full time, particularly parents and caregivers, are often left with what amounts to a hidden deficit once work, commuting, caregiving, and basic self-maintenance are subtracted from the day.[47]

UBI does not force anyone to work fewer hours. What it does is convert a specific category of activity from something people can only do when they happen to have the time and money to spare, into something people can choose to prioritize on purpose. Cooking a real meal, exercising, sleeping enough, or simply being present with aging parents or young children stops being a luxury squeezed into the margins of an exhausting week, and becomes an ordinary, available choice.

5.5 Working Still Makes Sense, on Different Terms

None of this argues against paid work. A person can still work part time purely for enjoyment, or to build additional income, without the underlying anxiety that currently accompanies almost every job: the fear of what happens if it disappears. That distinction, between working because a paycheck is the only thing standing between a person and disaster, and working because the work itself is worth doing, is the entire point of this chapter.

5.6 From Imposed Purpose to Chosen Purpose

Psychologists who study human motivation have identified three basic psychological needs that drive genuine, sustained motivation: autonomy, competence, and relatedness. According to Self Determination Theory, developed by psychologists Richard Ryan and Edward Deci and now one of the most extensively validated frameworks in the field, people are most intrinsically motivated, meaning motivated from within rather than by external pressure, when they have genuine choice over their actions, feel capable in what they are doing, and feel connected to other people.[48] A wage economy that forces employment as the only legitimate path to survival satisfies almost none of these conditions for a huge share of the population; people take whatever job will pay the bills, regardless of whether it offers autonomy, a sense of competence, or genuine connection.

UBI does not hand anyone a sense of purpose. What it does is remove the imposed purpose that survival currently forces on people, the requirement to organize one's entire life around whatever paid position is available, and replace it with the possibility of chosen purpose: work, in the broad sense this chapter has described, selected because it actually satisfies the need for autonomy, competence, and connection, rather than selected because the rent is due.

5.7 What About the People Who Choose to Do Nothing?

This question deserves a direct answer rather than a dismissal. Yes. Some people will receive UBI and choose not to work in any meaningful sense at all, paid or unpaid. That will happen, and I will not pretend otherwise.

The more useful question is a different one: is it morally preferable to force someone to work, purely on the suspicion that they might otherwise be unproductive? UBI guarantees the right to survive. It does not guarantee admiration, status, luxury, or a meaningful life, and it was never designed to. Those things remain something each individual has to build for themselves. Removing the threat of destitution as the mechanism that compels people into work is not the same as removing the value of work itself; it simply removes coercion as the reason people engage in it.

5.8 What UBI Can and Cannot Give a Person

UBI cannot give someone a meaningful life. No government program can, and this book makes no claim otherwise. What UBI can do is remove the fear of economic survival, and in doing so, hand people the freedom to decide, on their own terms, what their lives are going to mean. That is a smaller promise than a guarantee of fulfillment, and it is also a far more honest one.

References

1. Wikipedia. "Pre-industrial society." Accessed 2026.

2. Barnes, R. "Data-Intensive Supercomputing in the Cloud: Global Analytics for Satellite Imagery." arXiv, 2017.

3. Marketing Scoop. "How Fast is Technology Advancing in 2025? Breakthroughs, Impacts & Future Predictions." 2025.

4. Cleverfolks (Medium). "How fast has Tech advanced in the last 10 Years?" September 2025, citing McKinsey Technology Trends 2025 and Statista Tech Analysis 2024.

5. Statista. "Adoption speed: telephone vs. mobile phone," chart, citing Visual Capitalist data on 50-million-user milestones for telephone, air travel, television, internet, Facebook, WeChat, and Pokémon Go.

6. UBS, cited in Milmo, D. "ChatGPT reaches 100 million users two months after launch." The Guardian / Reuters, 2023.

7. Scribd. "ChatGPT Slides," comparative time-to-one-million-users data for ChatGPT, Instagram, Spotify, Facebook, Twitter, Airbnb, and Netflix.

8. McKinsey & Company. "McKinsey Quantum Technology Monitor 2026: A commercial tipping point." 2026.

9. World Economic Forum. "Future of Jobs Report 2025." 2025.

10. McKinsey Global Institute, cited in Final Round AI. "AI Job Displacement in 2025 and 2026: Which Jobs Are At Risk?" 2026.

11. Goldman Sachs research, cited in Forbes. "These Jobs Will Fall First As AI Takes Over The Workplace." April 2025.

12. Marchant, G. "The pacing problem," cited in Brennan Center for Justice, "Congress Must Keep Pace with AI." 2024.

13. Constangy, Brooks, Smith & Prophete, LLP. "The Pacing Problem: Can AI Legislation Keep Up With Emerging Technologies?" April 2025.

14. "Beyond Fairness Metrics: Roadblocks and Challenges for Ethical AI in Practice." arXiv, citing 50-year-old U.S. fair lending laws.

15. Wikipedia. "Minimum Income Standard." Accessed 2026, quoting the original UK MIS definition.

16. Joseph Rowntree Foundation. "A Minimum Income Standard for the United Kingdom in 2025." 2026.

17. Liu, S., and Su, Y. "The impact of the COVID-19 pandemic on the demand for density: Evidence from the U.S. housing market." ScienceDirect / Regional Science and Urban Economics, 2021.

18. City Journal. "How Remote Work Led to Big Cities' Decline." October 2025.

19. Renwick, T. J. "Basic Needs Budgets Revisited: Does the U.S. Consumer Price Index Overestimate the Changes in the Cost of Living for Low-Income Families?"

20. The Motley Fool. "Here's How Much It Costs to Raise a Child," citing USDA "Expenditures on Children by Families," 2015/2017 editions.

21. U.S. News & World Report. "This Is How Much It Costs to Raise a Child in 2025." August 2025.

22. Scope (UK disability charity). "Extra Costs" and "Disability Price Tag 2023: the extra cost of disability." 2023–

2026.

23. University of Manchester Research Explorer. "The disability price tag: Technical report."

24. "Exit, voice and loyalty in the family: findings from a basic income experiment." Socio-Economic Review, Oxford Academic, 2022, on the Manitoba Mincome experiment.

25. Stanford Basic Income Lab. "Visualizing Basic Income Research," citing the 2008–2009 Namibia basic income experiment.

26. Perkins, et al. "Analysing the impacts of Universal Basic Income in the changing world of work." Human Resource Management Journal, 2022.

27. Corinth, K., and Mayhew, H. "What Can We Learn from Guaranteed Basic Income Pilots in the United States? Evidence on Employment Effects." American Enterprise Institute working paper, 2026; summarized in The Daily Economy, March 2026.

28. Acemoglu, D., and Restrepo, P. "Automation and New Tasks: How Technology Displaces and Reinstates Labor." MIT Economics, 2019.

29. "The AI Layoff Trap." arXiv working paper, 2026, citing Autor, Chin, Salomons, and Seegmiller (2024) and Brynjolfsson et al. (2025).

30. University of Chicago Knowledge repository. "Price effects of the Alaska Permanent Fund Dividend," synthetic control study of Anchorage-area CPI.

31. Economic Possibility Initiative. "Since implementing the APF — Alaska's quasi universal basic income — has Alaska seen higher or lower inflation?"

32. Jones, D., and Marinescu, I. "The Labor Market Impacts of Universal and Permanent Cash Transfers: Evi-

dence from the Alaska Permanent Fund." NBER Working Paper No. 24312.

33. U.S. Social Security Administration. "Benefits Planner: Retirement — Retirement Age Calculator," and Congressional Research Service, "The Social Security Retirement Age: An Overview," on the 1983 Amendments' phased increase from 65 to 67.

34. Federal Reserve Bank of Chicago. "The Federal Reserve's Dual Mandate"; Federal Reserve Bank of St. Louis, "The Dual Mandate in Conflict: Balancing Current Tensions between Inflation and Employment," 2026.

35. RegisteredNursing.org. "Federal Investment in Nursing Education: What It Means in 2026," on the Nurse Corps Scholarship and Loan Repayment programs.

36. The Hechinger Report. "Amid a national shortage of nurses, nursing apprenticeships are beginning to offer a solution to the problem." 2025.

37. Bipartisan Policy Center. "2025 Social Security Trustees Report Explained," on the declining share of workers paying into Social Security relative to beneficiaries.

38. Fortune. "U.S. workers just took home their smallest share of capital since 1947, at least." January 2026, citing U.S. Bureau of Labor Statistics data.

39. Fortune. "The record gap between corporate profits and worker pay has an 'undercurrent of betrayal,' top economist warns." February 2026, citing KPMG chief economist Diane Swonk.

40. U.S. Social Security Administration. "Social Security Board of Trustees: Projection for Combined Trust Funds One Year Sooner than Last Year." Press release, June 2025.

41. Yahoo Finance / CBS News. "Bill Gates Wants To 'Tax The Robots' That Take Your Job." 2025, and "The backlash

against Bill Gates' call for a robot tax," CBS News.

42. The Hill. "Bill Gates's robot tax idea is flawed. Here's why." 2017, and CBS News coverage of Lawrence Summers's response to the proposal.

43. Tax Foundation. "OECD Pillar Two (Global Minimum Tax)," TaxEDU Glossary, and Moody's, "OECD Pillar Two Explained: Global Minimum Tax Policy & Compliance," 2026.

44. Health and Retirement Study, University of Michigan, cited in Lifeworks Advisors, "Purposeful Retirement: Are You Retiring FROM Something or TO Something?" 2026; and Journal of Occupational Health Psychology research on career centrality, cited in Mentors After Retirement, "The Retirement Identity Crisis: Why Smart People Struggle (And How to Fix It)." 2026.

45. National Partnership for Women and Families. "Americans' Unpaid Caregiving is Worth More Than 1.1 Trillion Dollars Each Year." 2026 analysis of American Time Use Survey data.

46. Stockton Economic Empowerment Demonstration (SEED). "About SEED" and independent evaluation results, University of Tennessee and University of Pennsylvania researchers, reported by CBS News, "Experiment in guaranteed income leads to more work and better health."

47. The Decision Lab. "Can We Afford to be Time Poor? The Tax of Time Poverty." and Living Income for Sustainable Economic Prosperity (LISEP), "The Working Poor: How Time Scarcity Halts the Pursuit of the American Dream." 2025; meal preparation figure from Supermarket Perimeter, "Americans report lack of time for cooking, enjoying meals."

48. Ryan, R. M., and Deci, E. L. "Self Determination Theory and the Facilitation of Intrinsic Motivation, Social Develop-

ment, and Well Being." American Psychologist, 2000; summarized by the American Psychological Association, "Self Determination Theory: A Quarter Century of Human Motivation Research."